GUIDE

TO THE

Primitive Breeds of Sheep and their Crosses

ON EXHIBITION AT THE ROYAL AGRICULTURAL
SOCIETY'S SHOW, BRISTOL

WITH NOTES ON THE MANAGEMENT OF

Park Sheep in England

AND THE POSSIBLE ADVANTAGES OF CROSSING
THEM WITH IMPROVED BREEDS

BY

HENRY JOHN ELWES, F.R.S.
COLESBORNE PARK, GLOUCESTERSHIRE

WITH NUMEROUS ILLUSTRATIONS

JUNE 1913

Copyright © 2013 Read Books Ltd.
This book is copyright and may not be
reproduced or copied in any way without
the express permission of the publisher in writing

British Library Cataloguing-in-Publication Data
A catalogue record for this book is available from the
British Library

Sheep Farming

Sheep (*Ovis aries*) are quadrupedal, ruminant mammals, typically kept as livestock. Like all ruminants, sheep are members of the order 'Artiodactyla', the even-toed ungulates. Although the name 'sheep' applies to many species in the genus *Ovis*, in everyday usage it almost always refers to *Ovis aries*. Numbering a little over one billion, domestic sheep are also the most numerous species of sheep. A male sheep is called a 'ram' and a female sheep is called a 'ewe'.

There are four general styles of sheep husbandry to serve the varied aspects of the sheep industry - and the needs of a particular shepherd. Commercial sheep operations supplying meat and wool are usually either 'range band flocks' or 'farm flocks.' Range band flocks are those with large numbers of sheep (often 1,000 to 1,500 ewes) cared for by a few full-time shepherds, sometimes requiring the shepherds to live with the sheep as they move through the pasture. The pasture, which must be large enough to accommodate the sheep, can be either fenced or open and sheepdogs as well as means of transport such as horses or motor vehicles are commonly required. As range band flocks move within a large area in which it would be difficult to supply a steady source of grain, and consequently almost all subsist on pasture alone. This style of sheep raising accounts for most of the sheep operations in the U.S., South America and Australia.

Farm flocks are slightly smaller than range bands, and are kept on a more confined, fenced pasture land. Farm flocks may also be a secondary population on a larger farm, used by farmers who raise a surplus of crops to finish market lambs on or those with untillable land they wish to exploit. However, farm flocks account for many farms focused on sheep as primary income in the U.K. and New Zealand (due to the more limited land available in comparison to other sheep-producing nations). The farm flock is a common style of flock management for those who wish to supplement grain feed for meat animals. An important corollary form of flock management to the aforementioned styles are 'specialized flocks', raising purebred sheep. Many commercial flocks, especially those producing sheep meat, utilize cross-bred animals and breeders raising purebred flocks provide stud stock for these operations, often simultaneously working to improve the breed and participate in showing.

The last type of sheep keeping is that of the 'hobbyist'. This type of flock is usually very small compared to commercial operations, and may be considered pets. Those hobby flocks, which are raised with production in mind, may be for subsistence purposes or to provide a very specialized product, such as wool for handspinners. Quite a few people, especially those who have emigrated to rural areas from urban or suburban enclaves, begin with hobby flocks or a 4-H lamb before eventually expanding to farm or range flocks.

Sheep breeds are often grouped based on how well they are suited to producing a certain type of breeding stock. Generally, sheep are thought to be either 'ewe breeds' or 'ram breeds.' Ewe breeds are those that are hardy, and have good reproductive and mothering capabilities – they are for replacing breeding ewes in standing flocks. Ram breeds are selected for rapid growth and carcass quality, and are mated with ewe breeds to produce meat lambs. Lowland and upland breeds are also crossed in this fashion, with the hardy hill ewes crossed with larger, fast-growing lowland rams to produce ewes called 'mules', which can then be crossed with meat-type rams to produce prime market lambs. Many breeds, especially rare or primitive ones, fall into no clear category. Sheep breeders look for such traits in their flocks as high wool quality, consistent muscle development, quick conception rate (for females), multiple births, and quick physical development.

Keeping sheep can often be a complex business, as sheep have many natural predators – coyotes in North America, Foxes in Europe and dingoes in Australia. Lambs in pasture are particularly vulnerable, frequently falling prey to crows, eagles and ravens. Consequently, many sheep are kept in barns, paddocks and pens, as well as merely out in the open. Freshly shorn hoggets (a young sheep of either sex from about 9 to 18 months of age) can be very susceptible to wet and windy weather however, and are frequently kept inside. This can become hard to organise for the crofter, and consequently most Australian farmers arrange for all the ewes in a mob to give birth (the *lambing season*) within a period of a few weeks. As ewes sometimes fail to bond with new-born lambs, especially after delivering twins or triplets, it is important to minimize disturbances during this period. In order to more closely manage the births, vaccinate lambs, and protect them from predators, shepherds will often have the ewes give birth in 'lambing sheds'; essentially a barn (sometimes a temporary structure erected in the pasture) with individual pens for each ewe and her offspring. In Norway most of the ewes are examined with ultrasound equipment to determine how many lambs will be born.

Sheep are an important part of the global agricultural economy. However, their once vital status has been largely replaced by other livestock species, especially the pig, chicken, and cow. China, Australia, India and Iran have the largest modern flocks, and serve both local and exportation needs for wool and mutton. Other countries such as New Zealand have smaller flocks but retain a large international economic impact due to their export of sheep products. Despite the falling demand and price for sheep products in many markets, sheep have distinct economic advantages when compared with other livestock. They do not require the expensive housing, such as that used in the intensive farming of chickens or pigs. They are an efficient use of land; roughly six sheep can be kept on the amount that would suffice for a single cow or horse, and sheep can also consume plants such as noxious weeds that most other animals will not touch, and produce more young at a faster rate. Also, in contrast to most livestock species, the cost of raising sheep is not necessarily tied to the price of feed crops such as grain, soybeans and corn. Combined with the lower cost of quality sheep, all these factors combine to equal a lower overhead for sheep producers, thus entailing a higher profitability potential for the small farmer.

Primitive Breeds of Sheep and their Crosses

MY attention was first attracted to the good qualities of what I may term the Primitive Breeds of Sheep, by the purchase, some years ago from a neighbour, of some black ewe lambs of Welsh origin, which I turned out in my deer park, and which were able to live on the same pasturage and under the same conditions as the fallow deer which have been there for over a century. I then imported a few ewes and a ram from the island of Soay, through the help of the proprietor, Macleod of Macleod, and crossed some of my black ewes with this ram, as well as with some Exmoor-bred Cheviots and Scotch Black-faced ewes. To these I have added a few of all the original breeds now surviving in Britain, and have made numerous experiments in inter-crossing them.

Though these experiments are as yet only in their infancy, I have no doubt that they will throw much light on many obscure points as to the inheritance of colour and horn on reversion ; and on the influence of environment on the wool, constitution, and fertility of sheep generally : and following the admirable example which Professor Cossar Ewart, F.R.S., Professor of Natural History at Edinburgh University, gave at the special request of the late King Edward, by his exhibition of Horse and Zebra hybrids at the Society's Exhibition at York in 1900 ; I have brought a selection of these sheep to Bristol, in the hope that they may prove interesting and instructive both to Agriculturists and Zoologists.

I do not propose in this guide to enter into a discussion of the much-disputed questions of Telegony, Saturation, Transmission of Acquired Characters, and Relative Influence of Male and Female Parents; a better knowledge of which are of such vast importance to breeders of pedigree stock; neither shall I as yet attempt to show how far my experiments bear out the Mendelian Theory of Breeding in sheep; because all these questions are very difficult and complex, and my opinions on them are as yet very uncertain. But as far as my opportunities permit, I intend to carry on the experiments, and to keep as I have done, exact records of the results obtained. I shall welcome the co-operation and assistance of any persons who are disposed to assist me by taking over one or more of the larger breeds, for the study of which my land of Colesborne is not good enough, and where I have not space to study properly so many different breeds and crosses simultaneously.

I have, perhaps, tried to do too many things at once; for which my excuse must be, that at sixty-seven one's expectation of life is too short to complete such a study if undertaken by successive steps; but I shall not have worked in vain if I succeed—to use the words of Professor J. Arthur Thomson—in making the science of breeding less empirical and the facts of inheritance less obscure.

The object of the exhibit is to show as far as possible in the space allotted, the primitive British breeds of sheep still existing; which are not seen at Agricultural Shows, but which have been preserved in a more or less pure state, and in some cases represent the stocks from which our modern improved breeds have been developed.

Agriculturists as a class will be disposed to think that on account of their small size, slow maturity, and more or less wild disposition, these sheep are not worth their attention; but I am able to show that if properly treated, they have many valuable qualities which are too often deficient in the breeds recognised by Societies; and that by judicious crossing they are able to transmit these qualities to more modern breeds.

After thirty-five years' experience of breeding sheep on

my own land in the Cotswold hills, and for ten years in Hampshire and Essex, I have had unusual advantages for observing a number of the pure breeds and of many different crosses between them. I have learnt that though on the better soils, the improved breeds of their respective districts are usually the most profitable on well-farmed land; yet that on the poorer soils, they degenerate unless artificially fed to an extent which is not profitable; and I believe that the modern tendency to over-high feeding, which must be practised by ram-breeders who have to compete with each other at Shows and Sales, is very prejudicial to the health and constitution of the stock bred from them.

A striking instance of the bad results of such practice is given by Professor Wallace in *The Farm Live Stock of Great Britain*, 4th edition, pp. 528-529 (1907), where in speaking of the Improved Cheviots bred by the late James Brydon between 1850 and 1880, which became the most fashionable and highly priced rams of the breed, he says:

"The mistaken practices of intense in-and-in breeding and overfeeding of mountain sheep were quite sufficient of themselves to accomplish the wreck, which entailed a widespread and heavy loss on the Cheviot sheep-breeders of the South of Scotland, and from which it is questioned if the breed has ever entirely recovered. Brydon began before 1860 the practice of housing the ram lambs during winter. The old rams went into the house when they came from the ewes. The sheep bred from Brydon's rams grew long and lanky, their coats were open and watery, and, worst of all, the ewes became very lean in winter, developed into bad milkers, and in consequence deserted their offspring, which, being naturally soft, made a very short struggle for existence. With all their bad qualities they were very prolific, as many as from half to two-thirds of the ewes on good hefts producing twins, half of which had to be killed in bad seasons.

"A few trying years of bad weather sufficed to show that in attaining size and beauty Brydon had got on the wrong tack and had sacrificed hardiness and utility."

If ram-breeders in England were always as honest as Professor Wallace, and could afford to tell of their misfortunes as well as of their successes, I venture to think that similar instances could be found in England. As an example of an exactly opposite system I will mention the practice of Mr. John Robson of Newton, one of the most successful breeders of Cheviot and Scotch Blackfaced sheep in this country.

When I found twenty-five years ago that I could not keep my native Cotswold breed on land which had been laid down or fallen down to grass, I looked about for a smaller and hardier breed, and bought of Mr. Robson 100 Cheviot two-year-old ewes which had been wintered hard on the Cheviot hills without roots, and were not bred from until they were a year older than usual. When I first saw them I certainly did not think they were worth anything like the price asked for them, but I took his word for it, and have never had ewes to equal them for wintering on poor grass, for suckling their lambs, or for keeping their teeth sound to a great age. To show the longevity of some of these unimproved breeds, I have brought a Hebridean ram which I believe to be 14 years old, whose teeth are still sound and who is the sire of several doubles this year.

The objects which I have had in view during the three years I have been making experiments, which I hope to continue for a considerable period, are as follows :

Firstly, to produce a breed capable of enduring the extremes of wet, cold, and heat which my district suffers from,[1] with a death-rate not exceeding 2 or 3 per cent from all causes.

Secondly, they must be able to winter on grass alone without suffering from lameness, to which all the improved sheep that I have tried, especially the Down breeds, seem to have an increasing, and probably hereditary, tendency.

Thirdly, to produce and suckle their lambs without the shelter of a ewe pen, and without assistance from the shepherd, which is often necessary in the improved breeds.

[1] I may say that in 1911 the drought was continuous for four months, and the extremes of temperature registered were 22° in June and 92° in July ; and whereas in 1912 the rainfall was 50.64 inches, the rainfall of 1st May 1912 to 1st May 1913 was 53.10 inches. I mention these figures to show the effect of climate on the fleeces shown with the sheep.

Fourthly, to get a fat lamb in July and August without any more artificial food than may be necessary to prevent the ewes from scouring in spring when the new grass comes; or, if the lambs are kept over winter, to make a small carcass of high-class mutton, not exceeding 40-50 pounds dead weight at 18-20 months old, off the grass.

Fifthly, to produce as far as possible, without sacrificing the carcass, a fleece of fine soft wool suitable for making such cloth as I am showing.

With regard to the first four points, I may say that though the sheep have been collected from many different places, in some of which they have had much better grass than at Colesborne, I have been very much pleased with the results.

With respect to the wool, I have not as yet had time to determine whether some of the wools (which will be found in the exhibit of the Royal Agricultural College of Cirencester in the tent devoted to Education) are more valuable for hosiery purposes than for cloth; but Professor Barker of the Technical College of Textile Industries at Bradford has kindly undertaken to make spinning tests of the various qualities, which he considers the best method of testing their quality and value for different purposes; and the result of these tests I hope to publish next year in the Journal of the Royal Agricultural Society.

The clip of 1912 has been spun undyed into yarn of the three natural colours—white, black, and brown—and made into cloth suitable for ladies' and children's wear by Messrs. P. & R. Sanderson of Galashiels. This cloth has been handed over to Messrs. Hamper & Fry of Cirencester, whose representative is in attendance to take orders for it. I wish to observe that softness, lightness of weight, and texture have been the objects in view, and that the cloth is not intended for winter shooting-coats.

Another point which I have had in view in collecting these sheep is the production of woolled skins, suitable for mats, and for use in lining or trimming winter coats. The skins of the Manx, Welsh, Soay, and Shetland sheep are well suited for the former purpose without dye, and by crossing with the Wensleydale, as Professor Ewart has already done, the wool is much improved in length and lustre. Some skins of the crosses are shown.

Professor Wallace has called attention to the possibility of producing in this country, and in some of our colonies, the very valuable skins which are imported from Russia and Central Asia under the name of Astrachan and Persian lamb skins. A costly attempt has been made in America to import and breed the sheep which produce these beautiful skins, and I have little doubt that we might succeed even better if we could obtain the assistance and co-operation of the Board of Agriculture. But the restrictions which exist on the importation of live sheep into Great Britain are at present so difficult to overcome, and the expense of importing sheep from such a great distance overland is so great, that I have not felt justified in undertaking it alone. I am, however, confident that if suitable land in the south of England was available for the experiment, we might obtain very valuable results; and as the subject is of even greater importance to Canada, Australia, and South Africa, I hope that some gentlemen who have pastoral interests in those countries, or who are inclined to co-operate in this matter, will communicate with me, or visit Colesborne, where they will be better able to appreciate the hard conditions under which my sheep have lived and thrived.

An advantage which all these sheep have is that they seem very much less subject to the blow-fly than ordinary sheep, and this, I think, is due to their being much less subject to scour, and perhaps also to the nature of their wool. Whatever the cause, it is an immense advantage to sheep kept in large parks which are only occasionally visited by the shepherd. And in addition to their other good qualities these sheep have an individual beauty and interest, which make them very attractive in parks and to persons who do not keep a large flock of sheep or a professional shepherd. For the guidance of such persons I reproduce parts of an article which I wrote for the *Year-Book of the Amateur Menagerie Club, 1913.*

"First it must be remembered that these sheep are

descended from breeds which have been driven out of better pastures to the higher mountains and remote islands of Wales, England, and Scotland, and do not naturally live on rich grass or in small enclosures. Therefore a large area of poor, dry, but sweet land, is much more congenial to them, just as it is to deer, than fertile land. I do not mean to say that they cannot be kept on fat land, but from what I have seen, I believe that 100 acres of land worth say 4s. an acre will suit such sheep much better than 10 acres worth 40s. an acre, and that they will thrive better if thinly stocked on land which is grazed by deer, cattle, or ponies as well. I have noticed in my own park, that they usually prefer the higher and more exposed part of it, where the grass is neglected by the deer in summer, and though such coarse grass is not likely to fatten them in summer, yet, the harder it is grazed in winter the better it becomes. When snow or severe frost comes they require some keep just as deer do, and I have found that my Black Welsh sheep will gnaw the bark of Ash, Elm, and Beech branches cut and laid down for them just as readily as deer will do; whilst if hay is given them it must be sweet and well made, and not musty rain-washed hay. Heather no doubt is as good winter food for park sheep as it is for deer or Scotch Black-faced sheep, but I have no personal experience of keeping these breeds on heather land. Gorse is also a good winter and spring fodder for such sheep, and I have seen them living so much on it in Wales, that their lips were quite sore from the spines. Seaweed is in Shetland, Orkney, and the Hebrides a great help to sheep in winter and spring, as it is to deer, and in all parks where sheep are kept rock-salt should always be placed in boxes for the sheep to lick.

"Early spring, just when the new grass is beginning to shoot, is the critical time both for old and young sheep wintered on grass, and a little hay chaff mixed with oats is a great help both to the tegs and the ewes, as it prevents them from scouring at the time when they are weakest.

"One of the most valuable qualities in these sheep is their freedom from scouring, which is no doubt one of the reasons why they are so seldom attacked by blow-flies. In most parts of England it is necessary to go round ordinary sheep once or even twice a day in hot weather, to look for fly-struck sheep, but some of the breeds which I propose to describe seem to be almost immune to maggots, though at present I do not know whether this is due to their wool or to some other peculiarity. Horned sheep are often much troubled in summer by small flies settling at the base of their horns, and causing them to rub their heads against rails or trees till they become sore, and if this is noticed some strong smelling oil, grease, or tar should be applied at once, or a piece of fresh tar line tied round the base of the horns.

"If they become very poor in winter, so that the growth of the wool is distinctly checked, as it is in the native Shetland sheep, it sometimes becomes loose, and sheds when they begin to thrive again in early summer; and in such cases it may perhaps be better to pluck the wool instead of shearing it, but this practice, though generally adopted by the Shetland crofters, should not be necessary as a general rule, and as the sheep thrive better after shearing when the weather gets warm, it is better not to leave them unshorn later than about the first or second week in June, though Cheviot and Black-faced Scotch ewes in their own country are often not shorn till the first week in July. In some parts of Wales the lambs are shorn about the end of June, and though they will no doubt grow faster if shorn than if carrying a heavy fleece during the hot weather, I have not adopted this practice. I do not think that it is worth while to wash this class of sheep before shearing if they have been kept on clean land, as whether washed or not the fleeces must be scoured by the manufacturers afterwards; and the difference in weight between washed and unwashed fleeces is not nearly as great as in the case of heavy breeds which have wintered on roots.

"A very marked feature in most of the mountain breeds is that the ewes are excellent mothers. It is rarely necessary to give the ewes any help in lambing, as the lambs get up and suck their own dams without any help, and however poor the ewes may be themselves, they usually have plenty of milk, so that the lambs grow fast as long as they are suckled. In large parks it seems doubtful whether it is better to wean the lambs or to let them suck as long as they will. If the wether lambs are to be sold, I should leave them unweaned till September or October, when they are usually worth more to kill than to winter, and sometimes at 3 or 4 months old are as heavy as their dams. But if they are weaned, both the ewes and the lambs must be kept in fences which they can neither climb, jump, nor get through, as until they settle down they are very difficult to keep in bounds.

"I must also say that one of the first and most important points to consider in keeping these sheep is the fences. Ordinary walls or hurdles they can jump or climb, unless there is a wire stretched a few inches above the top as usually done in the north of England. I have seen Shetland sheep jump a five-foot wall without hesitation. A good hedge, if not thin at the bottom, will stop them better than a wall, and they do not attempt to break out of wire fences or wire netting 3 or 4 feet high, but the tendency to break bounds is not great, unless acquired in youth or by bad example, and I have had much less trouble in this respect than I anticipated. The wild nature of these sheep is always shown among the rams by their uneasiness, and desire to fight during the rutting season, when if kept from the ewes they are difficult to keep in, and usually lose condition rapidly. If a strange ram is put with others they will attack him just as deer will, and I have had several killed or injured by fighting.[1] They retain their vigour to a great age. One of my Hebridean rams, believed to be 14 years of age, got out of the field where he was kept, by swimming over a good-sized stream to get at some Cheviot ewes on the other side, and this ram still has his teeth sound, and is in service.

"A Soay ewe, said to be 12 years old, bore a healthy lamb, and there are instances on record of Cheviot and Welsh ewes breeding at a greater age than this. In my experience

[1] My best five-year-old spotted ram has just had his horn broken off close to the head in fighting, March 16th, 1913.

these ewes are not so prolific as those of larger breeds, but this I think is due more to low feeding than to the breed. It is a common thing in the North of England for pure or cross-bred mountain ewes, which normally have only one lamb when wintered on the hills, to bring two and sometimes three lambs when thinly stocked on good grass-land in the valley; and if these ewes are helped with a little corn or cake at lambing time and afterwards, they will bring up two good fat lambs as easily as they would suckle one on mountain pasture. The presence of extra rudimentary teats on the udders of some sheep, a subject which has received but little attention, seems to prove that it might be possible to fix by selection a race capable of rearing three or four lambs, and some little-known Russian breeds are reported to produce three or four lambs normally; but in my opinion it is better to have one strong, well-nourished lamb than two less vigorous, especially under the conditions in which park sheep are kept.

"With regard to horns, little attention has in this country been paid to this character, except by breeders of Black-faced Scotch and Dorset sheep, who look on the shape, size, and position of the horns as an important point in the ram. They consider that the horns should be set wide apart on the head, and that the curl should be open and not close, because if the horn grows in a close curve it is apt, as the ram gets old, to grow into the side of the face. Among the four-horned Hebridean, Manx, and Spotted sheep, the primary horns sometimes—and the secondary lateral horns often—curl round as they grow in length, so that the point of the horn grows into the cheek or jaw; or it may curve down below or beyond the sheep's jaw, so that it cannot graze on short grass. Whenever such a tendency is seen, it is necessary to cut off a slice on the inside of the horn, or to cut off the end entirely, and as the appearance of the ram is spoilt by such shortening, it is better to select as ram lambs those with wide-set and spreading horns.

"It often happens that the secondary horns are not formed on a bony core, as normal sheep's horns are formed, but are loosely attached to the skin only, and these horns are easily knocked or rubbed off when the lambs are young, therefore some care is necessary when penning or handling them not to injure these young growing horns.

"Five horns are rarely produced by some breeds, but I have never seen a skull with more than four horn cores, and a Manx ram lamb which I purchased this year, which had a well-developed fifth horn on one side between the primary and the secondary horn, lost it when about seven months old."

I will now proceed to describe the various breeds shown, and the crosses from them.

The Old Horned Wiltshire Sheep

The Old Horned Wiltshire or "Western" sheep as Professor Wallace calls it, has been extinct in its native county for so long a period that the late Mr. Squarey of Salisbury, who was a very old man, had no distinct recollection of it. Low, in his folio published in 1842,[1] gives an excellent coloured picture of a ram and ewe of the breed, which then only survived in Wilts on a farm at Hindon bequeathed and held on the condition that the proprietor should keep a pure flock of this breed. I have a photograph from a picture in the possession of Mr. Rawlence of Salisbury, painted in 1809, which shows the exact type of animal which exists to-day, with the one exception that the horns were apparently rather shorter and the wool longer. One of the most remarkable features of the breed as we now know it is the fleece, described by Low as being only about $2\frac{1}{2}$ lbs. in weight, and bare on the belly. Now it is so very short, and sheds naturally so early in the year, that it is not thought worth the trouble of shearing, and drops off in May, or later if the sheep are very poor; the lambs even peeling when 3 or 4 months old as clean as though shorn.

Mr. Berry of Brampton Ash,[2] near Market Harborough, who has bred them since 1877, tells me that during all

---

[1] *The Breeds of Domestic Animals of the British Islands.*
[2] I am indebted to this gentleman for the photographs of his flock reproduced on Figs. 5 and 6.

those years he has selected those which have least wool and lose it earliest, because he has found by experience that the cross lambs which have least wool fatten the quickest. He tells me, further, that the breed has been kept pure in the Midland counties for nearly 100 years. But, notwithstanding its southern origin and absence of wool, this breed is as hardy as any sheep I know, and cares nothing for wet or cold. It is a very large-boned breed, with high withers and broad back if well bred, but sometimes narrow over the heart and wanting in symmetry. At some early period breeders of fat lambs in the Vale of Aylesbury, Northamptonshire, and Leicestershire discovered that for crossing with Welsh, Scotch, and Down ewes it had an extraordinary value; and in consequence a few small flocks have been maintained in a pure state—or in some cases slightly crossed with the Dorset breed—which latter at some early period was probably very nearly allied to, and perhaps only an improved off-set of, the old Wilts. Wallace tells us that Wilts × Hampshire lambs at 3 months old should weigh 6 to 8 stone dead weight; and Mr. C. H. Monk of Sheepcot Hill, Aylesbury, who is a leading breeder, tells me that the cross with Welsh ewes is the most profitable, if brought out in August and September, when they weigh 4 to 5 stone dead weight.

When visiting Mr. Monk's flock in 1911, I purchased from him a white ewe with two lambs, which was bred by him from a pure Soay ewe, imported from St. Kilda, and a horned Wilts ram of about five times her weight. One of these is now shown as a wether (Fig. 1), the other is a ewe which, when put to a Piebald ram (Fig. 2), produced in 1912 the twins shown in Fig. 3. The same ewe put to a pure Wilts ram lamb has this year produced twins (Fig. 4), which like their dam show no trace of Soay parentage except in their smaller size. Nothing can more strongly illustrate the prepotency of the old Wilts blood than these crosses, and it is even more remarkable that in form and condition the produce are distinctly superior to both of their parents, and for hardiness and as good mothers are equal to any sheep I possess.

The only place where these rams are publicly shown is at Northampton Ram Fair in September, and the number shown is quite small, as the majority of the ram lambs are sold privately. The only other district I know where these rams are used is in North Wales, where, as I am told by Mr. Chambers of Castellior Farm, Anglesea, they were introduced about thirty-five years ago, and are now much used for producing fat lambs from Welsh ewes. He told me that for early lambs the Southdown ram was preferred by most breeders, as the quality was better and they fatted more quickly; but in July and August, when there is a good demand for lamb in most counties and North Wales, the Wilts cross was larger and more profitable.

The only published account I know of experiments in this cross is in a series of papers by the late Professor Winter, made by him between 1901 and 1912, on the Madryn and Aber Farms of the University College of North Wales at Bangor; and the general result of these experiments, which are given with full detail as regards age and weight, but without any prices or profit and loss figures, seems to show that the Southdown cross is in some seasons the better one; but this no doubt depends on the quality of the grass and the amount of artificial food given, and perhaps on the merits of the individual parents as much as on the breed, and therefore, to my mind, is not conclusive.

It is said by some that if the Wilts ram is used, the lamb must be fattened off the ewe or not at all, as the cross is very slow to fatten when older; but this is not the experience of Mr. Monk, who assures me that he has seen a Wilts and a Down ram run together with a flock of ewes, and the lambs kept for tegs, in which case the Wilts cross lambs were fat before the others.

I saw his flock again in May 1913, and found them wintered on grass-land without roots, in better condition than most Down flocks; and notwithstanding an unusually wet winter and spring, there was no sign of lameness. Some of the ewe tegs were suckling young lambs, and seemed but little inferior to the older ewes in size and condition. But it seemed to me that to have these sheep in good condition they must be kept in small numbers on fairly good grassland, and that they are not so tame or easily managed as sheep which have been brought up in hurdles. The old rams are sometimes inclined to be vicious, and are very strong and active in comparison with other heavy sheep. To illustrate this, I was told by Mr. Monk that his father had once run a Wilts ram with 100 Oxford ewes, 69 of which produced twins, and that after this the same ram had served 100 Welsh ewes during the same autumn.

The Norfolk Sheep

The earliest account I know of the Norfolk sheep is in Marshall's *Rural Economy of Norfolk* (1787); he says that at that period sheep were so scarce in most parts of Norfolk that on a ride from Thetford through Walton, Dereham, and Reepham to Gunton, he did not see one sheep until he got within a few miles of the end of his journey. There were then two varieties; one larger, weighing from 15 to 25 lbs. a quarter, which was the common stock of the county; the other smaller, from 10 to 15 lbs. a quarter, mostly kept on the heaths near Brandon and Methwold; these latter went by the name of "heath sheep," but differed in no respect from the common sort except in being smaller and their wool being finer. The characters of the Norfolk sheep in those days were:—The carcass long and slender; this is true to-day. The fleece short and fine; this is also true to-day, as a fleece given to me by Mr. Geoffrey F. Buxton of Dunston was considered, when examined with other fleeces by experts at Galashiels, to be equal to the very highest-class Cheviot from Caithness, and worth about 2d. a pound more than that of the Cheviots of the border counties. The legs long and black or mottled. The face black or mottled (the mottling seems to have nearly disappeared at the present time, all that I have seen having pure black faces and legs). The horns of the ewes and wethers resembling those of the Dorset ewes, but those of the rams very large, long, and spiral like the horns of the Wiltshire ram; this is also the case to-day. The sheep then, as now, seem to have been deficient in their fore-quarters, narrow over the heart, and with a high and narrow chine. They were very slow feeders as compared with the Suffolk breed, which has undoubtedly descended from them. Marshall says: "They may be bred and will thrive on heath and barren sheep-walks where nine-tenths of the breeds in the kingdom would starve, they stand the fold perfectly, will fat freely at two years old and bear the drive remarkably well to Smithfield or other distant markets, and the superior flavour of the Norfolk mutton is universally acknowledged." I may say in confirmation of the above remarks that my grandfather, who kept these sheep on Congham heath eighty years ago, used to say that they would travel on their own legs to Smithfield market, nearly 100 miles, in a week without suffering; and that the dressed carcass of a wether which I saw hanging in the Earl of Leicester's larder weighed 93 pounds off the grass, and the meat is considered by him to be as much like venison as mutton in flavour.

When Low wrote in 1842, the pure breed was already becoming rare, and had been much crossed with Leicester and Southdown rams, and according to a paper written in 1847, quoted by Wallace, the Southdown cross was then general, and the parents of the Suffolk breed; though that name was not officially recognised until about 1859. The wonderful success of this breed as a butcher's sheep is best proved by the fact that there was hardly a cross-bred carcass in the prize list of the Butchers' Classes at the last Smithfield show which had not Suffolk blood in it.

The number of Norfolk sheep now surviving in a pure state is very small, but it is sincerely to be hoped that the landowners of their native county will institute a class for them at the County Show. The value of the breed to-day is rather for crossing than in a pure state; but this has been proved by Mr. McCalmont, who won the first prize in the Carcass Competition for cross-bred sheep at Smithfield in 1912, with a wether by a Southdown ram out of a Norfolk ewe, which weighed 102 pounds dead weight at 21 months old. This cross is nearly always hornless, as I am informed by Mr. McCalmont's farm manager; but I expect that if a cross were made between a Norfolk ram and Cheviot or Welsh ewes, the produce would be strongly horned. The principal objection to these sheep is that, being very strong and active, and of a less placid disposition than most sheep, they are slow feeders and difficult to keep in ordinary fences; the old rams are, at times, vicious, and so strong that they can knock a man down.

The Shetland Sheep

As the wool of this breed is the finest and most valuable of any wool grown in Britain, I shall quote at some length from an article I wrote in the *Scottish Naturalist*, January 1912.

"According to Youatt and other writers, the Shetland sheep were originally of Danish or Scandinavian origin, but little if any reliable information seems available on this point; most accounts seem to be based rather on hearsay than on personal knowledge or research.

"An exception is, however, to be found in Dr. L. Edmonston's *General Observations on Shetland* (1840), who said: 'The sheep is small, not often horned, ears pointed and erect, face, back, and tail short, fine-boned, legs long; naturally wild, active, and hardy, and little liable to disease. The colour generally white, sometimes ferruginous, grey, black, or piebald; the wool very soft and often fine. The more damp and moory the pasture, the softer is the wool; one of the causes of which probably is deficient strength and nourishment, another is the astringent nature of the food. A serious casualty affecting the value of a Shetland flock arises from the constant vicinity of precipices facing the sea; and great losses by their falling over the rocks are often sustained.

"'No breed can, as a rule, be better adapted to the Shetlands, than those that are native in them, and as they are always in demand, we should do well zealously to cultivate them. All that is necessary is such a sufficiency of food and care as will not encroach too closely on their habits and hardihood, and a persevering selection of the best animals for breeders; yet if premiums had been offered for producing change and degeneracy, it is difficult to imagine a course better calculated to produce them than that which has usually been pursued.'

"Of the truth of this latter statement I had ample evidence when I visited Shetland in 1911, for, except in some parts of Wales, I have never seen sheep so neglected as on the common grazings of the Shetlands.

"H. Evershed, who published a good paper on the agriculture of Shetland in the *Highland Societies' Transactions* about thirty-five years ago, states that the Black-faced breed was able to live and thrive wherever the native sheep could live, and owing to the much higher price of their lambs and the greater quantity of their wool (which at that time was worth twice as much per pound as it now is), they, together with Cheviots, half-breds, and cross-breds, had supplanted the native breed on all the improved land; leaving the native breed only in crofters' hands, on the very worst of the land and common grazings. He quotes Shireff for the fact that the native breed had been much mixed with Dutch sheep, during the time when the fishing was in the hands of Dutch merchants.

"He calls it 'a straight-horned or goat-like breed, the fleece of wool and hair mixed, weight not over two pounds. Its softness and fineness need not be enlarged on.'

"Prof. Wallace, in *Farm Live Stock* (1907), quotes the late Q. M. Hamilton, who said that 'Youatt's description does not hold quite good for the Shetland sheep of the present day, as the only two islands on which they are really pure are Foula and Papa Stour.'

"I was not able to visit either of these islands, but have purchased some of the best ewes from Foula this year, which differ in no respect from the light brown sheep (this colour is known as *murret* or *moorit* in Shetland) which I saw in several places, and though the wool of this colour is the most highly priced, on account of its use for shawl-knitting, it does not seem so fine or soft as some of the white wool of North Maven, neither does it approach in fineness the wool of two specimens of Shetland sheep from Unst, presented in 1871 by T. Edmonston of Balta Sound to the Edinburgh (Royal Scottish) Museum, where they are now exhibited—these are apparently the only specimens of the breed in any museum.

"These specimens consist of a hornless ewe, and what I believe to be a wether, with short horns ($6\frac{1}{2}$ inches long), of the usual wether type, the wool pure white and very fine, $2\frac{1}{2}$ to 3 inches long on the shoulder. The height of these sheep as stuffed is 22 to 23 inches, the length of body (breast to tail) 26 to 28 inches, the tail very short and broad at the base. This form of tail is considered typical of the breed. I estimated the weight of these sheep, if fairly fat, to have been about 30 lbs. dressed. But the weight of Shetland lambs, when really well fed, at four to five months old, is said to be sometimes as much as this; and in all these unimproved breeds the growth of the lambs, as long as they are sucking, is remarkably rapid during summer.

"As to the sheep of Papa Stour, I only know them from a very diminutive animal, which is stuffed, in the Domestic Animals Gallery of the Natural History Museum in London, labelled as from 'Papa, Orkney Islands'; but I am assured by Mr. Gerrard, from whom this specimen was procured, that its real habitat was Papa Stour, Shetland. If adult, as it seems to be, this is the smallest sheep I ever saw, and looks more like a freak than a distinct variety.

"During my visit to the islands I saw and learned a good deal about the sheep which now exist, and which probably have without exception some foreign blood, though the hard conditions under which they live tends no doubt to the survival of those which have most Shetland blood in them.

"With few exceptions they are kept on the worst lands only, and as the grazings are common to a number of crofters, most of whom are as much fishermen as farmers, there is no selection of rams, some of which remain the whole year wild in the cliffs. About the end of May drives are organised in order to collect as many sheep as possible for their wool, which at that time is beginning to shed. But as it will not all come off at the same time, the sheep are gathered at intervals of a week or ten days, when the weather is dry, into stone enclosures, when each crofter plucks as much as will come off without force from his own sheep. This practice entails a great deal of hunting with dogs, which must be very injurious to the weak ewes and their lambs. In the beginning of June I saw many sheep which had lost part of their fleece, and a great deal of shed wool was scattered about the hills. It seemed to me that the practice of plucking has been kept up, because there is a distinct break in the growth of the wool, similar to that which takes place in England when sheep have been ailing or starved; and that when the new wool begins to grow again in the spring it pushes up among the old wool. In the majority of the sheep that I examined I could not see any distinct difference between the hairy outer wool and the fine wool beneath, such as is described by some writers; and neither in dressed skins which I bought at Lerwick, nor in Shetland sheep which I have kept in England, have I found evidence of this difference. The grey coloured sheep (here called 'Sheila') seem to have a much longer and coarser fleece, as though crossed with the Black-faced breed, and neither the white nor the black sheep bred in England had fleeces as soft as they are in Shetland. I also found a good deal of kemp in the breech and hind parts of some of the fleeces I examined.

"The fine-wool spinners in Unst informed me that as they only require a few ounces of wool for the best quality of shawls, which are worth several pounds, they select only a little of the finest wool from the neck and shoulders, and that for this purpose it was better in North Maven than in Unst or Yell.

"This seems to be borne out by the sheep I saw at Lochend, where I bought a half-bred Cheviot, whose fleece was superior in quality, and more than twice as heavy as the fleece of some of the nearly pure Shetland sheep which I got from Mr. Gordon in Mid Yell.

"Mrs. Bruce of Sumburgh, owner of Fair Isle, who keeps a small flock of pure Shetlands and does much to encourage the knitting industry, tells me that it is not necessary to pluck the wool of her sheep, which, however, are much better fed than most crofters' sheep; and I am informed by Mr. Kerr, who has charge of a considerable number of Shetland sheep belonging to Mr. Stephens in Wiltshire, that he obtains 2s. a pound for shorn and washed fleeces of moorit-coloured sheep wintered on grass in that county.

"With regard to horns, I am not able to say what is the best or the true type of horns in Shetland sheep. Often the rams have none (Fig. 7), and these are preferred for their wool by some breeders. Many have short horns; but in a white ram which I exhibit the horns form a complete circle, as in Fig. 8. Some ewes have short, curved horns, but the majority have none, and I never saw any with straight or goat-like horns. A cross with the Black-face produces strong horns, and in some cases four are found in the rams of this cross. I bought in Mid Yell a ram with four horns (Fig. 9), out of a grey hornless ewe by a Black-faced tup which had again produced a four-horned ram lamb in 1911. I also saw in Mr. Haldane's house at Lochend, and at Mr. Anderson's of Hillswick, stuffed heads with four horns, very similar to the head of the Iceland sheep shown in Fig. 11."

Since writing this I have had evidence of the virtues of the breed: first, in Cumberland, where Mr. Howard of Greystoke was good enough to winter the ewes which I imported; secondly, at Penicuik, Midlothian, where Mr. Cowan has kept a small flock with great satisfaction for four years; thirdly, at Colesborne, where they do as well as any sheep I have tried; and lastly, on a farm on the Wiltshire Downs where Mr. Stephens has a flock which by care and selection have become very uniform in type and colour. The latter, like Mr. Cowan and myself, has had the wool made up into cloth, which is infinitely superior to the loosely woven and badly wearing material usually sold as Shetland homespun; and though I do not think that in the south the wool will ever be as suitable for fine spinning and hosiery as that grown in the islands, yet for my own purpose, namely, a light, soft, and warm cloth, suitable for ladies' and children's wear in spring and autumn, and for light summer overcoats, its softness and colour are all that can be desired.

The moorit colour,[1] which is found in only a small proportion of the sheep I saw in Shetland, seems to be easily fixed by selection, and I did not have a single white, black, or spotted lamb from any of the twenty-five ewes I bought, though some of them have a little white on the head, tail, and feet. The rich chocolate brown of the lambs tends to fade as the sheep get older and the wool grows, and becomes somewhat bleached on the outside by exposure to sun and rain, but it does not fade when made up into cloth as much as the black Welsh or Hebridean wool.

Mr. Cowan prefers the black and grey (sheila) colours, which are also wanted in the islands to make the borders of knitted shawls, but I intend to breed as Mr. Stephens has done, mainly from the brown or moorit-coloured sheep, and have no doubt that the price of wool of this colour will increase when sufficient quantity can be produced to attract the attention of fancy tweed makers. As far as I can see it does not tend to become much longer, or harsher, in the south, if the ewes are kept under natural conditions on poor pasture, but it will probably be wise to import rams from time to time to correct this tendency when it appears.

A cross between a white Shetland ram and a few Herdwick ewes (Fig. 10), made at my suggestion by Mr. Howard at Greystoke, has been successful in improving the harsh and coarse character of the wool of the Herdwick, which is about the lowest-priced wool produced in this country. The value of this cross-bred wool for cloth-making I have not yet had time to test, but intend to breed more, as the report on the fleece shown is so satisfactory.

A cross between the brown Manx and the Shetland has also been tried, and though the quantity of the wool and the size of the sheep is increased, I do not think the cross is so good as one between the Cheviot and Shetland, which, so far as I can judge at present, improves the carcass immensely, without losing much of the softness of the fleece ; and in making this cross again I shall choose the very finest-woolled Cheviot that I can procure from Caithness, where the

---

[1] This term is probably derived from a Scandinavian word meaning "moor red."

wool is considered to be worth more than that grown in the south of Scotland.

Mr. C. M. Douglas of Auchlochan has tried this cross with success, and the tweeds made from his wool by Messrs. P. & R. Sanderson are of very superior quality.

THE MANX SHEEP

The Manx breed of sheep exists in such small numbers in its native island that, but for the interest which has been taken in it by one or two residents, it might have become quite extinct. It was described 100 years ago by Parkinson as one of the smallest breeds known to him, averaging only 20 lbs. dead weight, with a maximum of 32 lbs. at three years old. This seems to show that the sheep were then, as now, restricted to the poorer hilltops of the island, where larger sheep could not live; for during the three years I have had them, though they have lived on poor grass alone, they have increased in size, and still more in the amount of wool which they produce. Youatt spoke of them as bearing much resemblance to Welsh sheep, being both horned and polled, and mostly of a white colour, with most of the peculiarities and bad points of the latter breed; but it seems doubtful if he had ever seen them, as to my eye they have much more resemblance to the Shetland breed, especially when of the favourite brown colour called "loaghton" in the island. White and black sheep are found amongst them, but I never saw a pure Manx sheep without horns, and in the rams four horns are more constant than in any other breed I know, possibly owing to selection. The old ram I show (Fig. 12) is a very good specimen of the breed, and the yearling ram, sired by him (Fig. 13), bids fair to be even better in horns, body, and wool. The white Manx ram (Fig. 14) is also a nice sheep.

Four horns are sometimes found among the ewes as well, but I consider the ewe shown (Fig. 15) to have the right type of horn. Professor Wallace, in his *Farm Live Stock of Great Britain*, gives an interesting account of the crossing of a Manx ram with Black-faced Scotch ewes, which proves that the four-horned character is very prepotent in this breed, and two of the heads figured by him are so like the heads of Hebridean rams, that I should not have been able to distinguish them.

A brown Manx ram (Fig. 17) which I lent to the Hon. M. Hicks Beach, M.P., put to a white Welsh ewe, produced the brown lamb shown with its dam.

I have bred a five-horned ram lamb this season (Fig. 15), but I am doubtful whether the fifth horn has a bony core or is, like one I had before, attached to the skull only by the skin.

The principal defect of the Manx breed, so far as my three years' experience of them enables me to judge, is that, probably owing to being very much inbred, they are not so hardy or such good mothers as the Shetland or Welsh sheep; and this defect I hope to cure by a cross with a Shetland of similar colour, which also gives a finer and more valuable, if a lighter fleece.

I am indebted to Mr. J. C. Bacon of Santon for much information about the breed, and for showing me three of the four small flocks which exist in the island.

The Soay Sheep

This is the smallest aboriginal sheep now known to exist as a pure breed, and is so nearly allied to the wild Moufflon, found only in the islands of Corsica and Sardinia, that Professor Ewart finds no marked difference in its skeleton, horns, or throat fringe.

It is a British representative of the short-tailed breeds found in Northern Europe, Iceland, and the Faroe Islands, which by some naturalists have been considered to have sprung from a different origin to the long-tailed domestic breeds, whose origin is quite unknown.

It is believed to have been the original breed of the island of St. Kilda, but is now confined to the small uninhabited islet of Soay, one of the St. Kilda group, where it exists in a state of absolute wildness, unseen by man except on the one or two annual visits of the St. Kildans to the islet in order to collect the eggs of the sea-birds which breed there in great numbers.

In my account of the Soay sheep in the *Scottish Naturalist* for February 1912 (Fig. 18), I said that the original breed of St. Kilda may or may not have been the same as those now on Soay; but there have been no four-horned sheep on the islands of St. Kilda in recent times, so far as known to Donald Ferguson, who has been ground officer of the island for twenty years, and succeeded his father in the office. But in Macaulay's *History of St. Kilda*, published in 1764, it is said that there were then about 1000 sheep on the main island, 400 on Boreray, and 500 on Soay, the latter alone being the property of the steward. He says: "Every one of those sheep had two horns, and many of them four. They are wonderfully fruitful. One of the people assured me that in the course of 13 months one sheep had increased his flock with nine more. She had brought forth three lambs in the month of March, and three more in the same month the year after, and each of the first three had a young one before they were 13 months old."

He speaks of the difficulty of catching the sheep, owing to the steepness of the rocks, and says that "the old rams, if chased into dangerous places and heated into a passion, turn sometimes desperately fierce; reduced to the necessity of yielding or tumbling into the sea, they face about and attack the pursuers."

As Mr. Millais, in his *Mammals of Great Britain*, truly says, it is quite distinct from the Hebridean four-horned breed, usually called "St. Kilda sheep," in English parks, and has all the habits and appearance of a wild sheep. Those which I imported three years ago from Soay were very wild at first, but when kept in a small enclosure, where people were constantly passing, in company with other sheep, they soon became as tame as domestic sheep; but some of these which I sent to Woburn last year, and which were kept alone and visited only occasionally, soon became as wild as ever, and would not let me come within 100 yards of them; and in Mr. E. E. Barclay's park at Brent Pelham, where they live and interbreed freely with the pure Moufflon, they are so wild that, though hounds are daily exercised in the park, they fly from the approach of a visitor.

The persistence of hair in the Moufflon and of wool in the Soay breed when living under similar conditions in this park is very marked, and in the crosses between the two breeds the tendency seems to be to revert to the Moufflon; but a cross which I saw in Paris, said to be between a Moufflon ram and a white ewe of the Saintonges breed, the lamb was black with a white belly and white patch on the head, and a very short woolly coat, and the tail was intermediate in length between those of the parents.

Owing to the extreme difficulty of landing on the island of Soay, I have been unable to see these sheep in their native home, and Mr. Eagle Clarke, who spent some weeks at St. Kilda, was also unable to land there, so that the information we have as to their number and conditions is mainly derived from a Gaelic-speaking native of St. Kilda. There seem to be two strains, in one of which the ewes have short horns, and in the other are hornless.

It is said that at some former time rams from St. Kilda were put on the island, and it may be from this cross that the white faces which occasionally appear among lambs bred from imported stock are due; but the true colour is a pale brown or fawn, with paler belly and mouth, and the lambs when first dropped are of a rich reddish fawn.

The breed is very long lived and prolific, two lambs being sometimes produced when the ewes are only a year old; but long interbreeding seems to have degenerated the race and made them smaller, and they do not thrive on my poor grass land as well as Shetland or Welsh sheep, and the very hot dry summer of 1911 seemed very trying to them; to do them justice they seem to require good feeding and plenty of room.

The breed would therefore seem to have little, if any, value from an agricultural point of view, for though their wool is very soft and fine, it is so short, and so apt to fall off in patches as the new wool grows in spring, that they are hardly worth shearing, the fleeces averaging under a pound in weight.

Mr. Lydekker states that it makes beautiful cloth, but some which I sent to be spun and woven by skilled hands in Argyllshire, under the direction of Mrs. Malcolm of Poltalloch, was unfavourably reported on, and the cloth made from it very inferior to what I have had made from other Scotch wools. I should not therefore have introduced them to the notice of agriculturists but for the crosses which I have bred from them, some of which seem to show little trace of the Soay blood, except in their smaller size, much finer wool, and greater hardiness; and I believe that these crosses if interbred, or crossed a second time with one of the parent breeds, may prove to have considerable value in improving the vigour and wool and reducing the size of breeds like the Southdown, Oxford, or Hampshire.

The first-cross with the Old Wilts (Figs. 3 and 4) is one of the most thrifty little sheep I have ever possessed; the ewes always having twins, which they nourish extremely well, and which seem to endure wet and cold and to keep their condition on poor land better than either of the parent breeds.

The first-cross with the Southdown is also an excellent one, the lambs fattening readily, and these first-crosses as interbred by Prof. Ewart show that the Southdown is the dominant blood, and that by careful selection it will be possible to fix a type of miniature, hardy, fine-woolled sheep, capable of thriving under conditions which the pure Southdown cannot endure.

Mr. H. Sanderson of Galashiels, in a letter to Prof. Ewart, says: "The more I look at the Soay × Southdown wool, the more I am convinced that it is a splendid breed to cultivate. It is distinct from anything we are used to, but of its fineness and handle there can be no question, and it is a big advance on all the wools at present grown in this country."

The cross between the Piebald ram and the Wilts and Soay ewe has also been very successful, though I have as yet not had time to develop it.

The cross between the Soay ram and Oxford ewe has also produced some excellent lambs, which show more of the size and character of the female parent, and the wool is distinctly superior in quality and not much less in quantity.

THE HEBRIDEAN SHEEP

The Hebridean breed is perhaps more generally known under the name of St. Kilda sheep, several flocks having been kept under that name in English parks for a long period, but my investigations lead me to believe that they are not the pure old St. Kilda breed, now preserved only on the rocky islet of Soay, but a cross from some primitive breed which existed in the Hebrides, and which may have been once introduced from Norway, more or less mixed with the blood of the Scotch Black-faced. I quote as follows from my paper in the *Scottish Naturalist*, 1912:

"Most writers agree in supposing that the old Hebridean breed was originally introduced from Norway, and some have supposed that the wrecked ships of the Spanish Armada may have brought some new blood, but we have no reliable evidence on this point. I can get no exact description, specimen, or figure of the aboriginal sheep which are said to exist on some of the islands of southern Norway,[1] and it is probable, indeed almost certain, that the breed now generally known as St. Kilda in English parks is a mongrel in which the Black-faced Scotch has a considerable share.

"Early writers give little exact information, but in Martin's *Description of the Western Islands*, 1703, p. 48, I find the following: 'About a league to the north of Pabbay, lies the isle of Sellay, a mile in circumference, that yields extraordinary pasturage for sheep, so that they become fat very soon. They have the biggest horns that ever I saw on sheep.' On p. 286 he says of St. Kilda: 'The horses and cows are of a lower stature than on the adjacent isles, but the sheep differ only in the bigness of their horns, which are very long.' He says nothing about the sheep on Soay, or about four-horned sheep.

[1] Dr. J. Walker, *Economic History of the Hebrides*, ii. 69 (1812), says, but I know not on what authority, that this native sheep of the Hebrides and the Shetland Isles is exactly the same with what subsists to-day in the Kingdom of Norway.

"Walker[1] says: 'The Hebridean sheep is the smallest animal of the kind. It is of a thin, lank shape, and has short, straight horns. The face and legs are white, the tail extremely short, and the wool of various colours; for besides black and white, it is sometimes of a bluish grey colour, at other times brown, and sometimes of a deep russet, and frequently an individual is blotched with two or three of these different colours. In some of the low islands, where the pasture answers, the wool of this small sheep is of the finest kind, and the same with that of Shetland. In the mountainous islands, the animal is found of the smallest size, with coarser wool, and with this very remarkable character, that it has often four, and sometimes even six horns.'

"How much Walker really knew about the Hebridean sheep of his time it is hard to say, but there is no evidence of the existence of sheep with short, *straight* horns anywhere in Scotland at the present day; and from what we know about the introduction, not only of Black-faces, but also of Spanish Merinos, which about this period were introduced into many parts of Scotland, it is probable that, except in the more remote islands, crossing had already taken place.

"Harvie-Brown and Buckley, in their *Fauna of the Outer Hebrides*, say very little about the sheep, except that four-horned sheep were not uncommon in Harris and North Uist, and that they had heard that even six-horned animals were not unusual. But I cannot find a specimen in any museum or collection to-day, except what have been bred in parks; and though I have myself seen in an island in West Loch Tarbert, Harris, in 1868, a ram with a fifth horn standing a foot or more high between the others, I cannot hear of any such alive at the present time.

"Millais figured, in *Mammals of Great Britain*, vol. iii. p. 212, fig. 4, the head of a Hebridean ram which has two horns, very much of the same type as those of my Soay rams, but longer, more spreading, and not so thick at the base. They are not at all like those of the so-called St. Kilda sheep, nor are they remarkable for their size. He informs me that he bought the animal alive in N. Uist from

[1] *History of the Hebrides*, ii. 69 (1812).

a crofter, and had to shoot it, as it was very wild. I have seen very similar horns on the old Norfolk Black-faced ram, and such a head might very well represent a primitive race from which the improved Black-faced Scotch sheep has been derived.

"Mr. A. M'Elfrish of Lochmaddy, in answer to my inquiries, writes as follows: 'I am afraid the subject is a pretty obscure one. There is certainly at the present time no such thing as a pure breed of four-horned sheep in these islands, nor has there been, since I came here in 1886, any such thing. No doubt there is a strain of four-horned blood running through great numbers of the sheep in these parts, but I know no one who will assert that he has a breed of such animals. My opinion is that all the four-horned ones that now crop up or that have cropped up in recent years are simply throw-backs. It is said that at one time, long ago, all the sheep in these parts were four-horned, and that they were replaced by the Black-faced from the Borders, but it would require some research to prove that. It is yearly becoming more and more difficult to procure a good specimen of a four-horned ram; but I proved one thing, at least to my own satisfaction, namely, that four-horned rams are certain, or at least almost certain, to throw four-horned lambs. A number of years ago I purchased from different parts of these islands a number of four-horned rams and put them to ewes of various kinds, native, crosses, half-Cheviots, etc., and in every single case without exception the tup lambs were four-horned, and in every case the horns were exact replicas of the horns of the sires; so much so, that any one could easily point out each ram's get. The one I think you refer to (Fig. 16) was a get of one of these, and was an exact replica of his sire, with the exception that by good grazing and a little hand-feeding in winter his horns developed enormously. The top ones at one time, I remember, measured on the tape 36 inches, and, as you say, the lower ones would very soon have prevented his feeding. They all but did so when he disappeared, but by grazing at the sides of slopes, banks, and ditches, he was able to pick up a living. At first he had five horns, the fifth growing from the centre of his forehead, but it was only skin deep and was early knocked off.'"

The breed is evidently impure because there is no fixed type, and though the late Mr. J. Macdonald of Balranald in North Uist took some pains to select the four-horned type from which the old ram I show (Fig. 19) is directly descended, no one in the Hebrides seems to have paid much attention to them since his death, and most if not all of the flocks in England have been crossed at some time, with small black sheep of Welsh, Breton, or other breeds. In some cases, as in the flock of Mr. Leopold de Rothschild, who has a large number at Ascott, near Leighton Buzzard, they have been kept mainly for their meat, in others they have run wild like deer in large parks without any attention, and have degenerated in horns, wool, and carcass.[1] Except for their fine horns and extreme hardiness, they seem to have no special value, as their wool is too long and coarse and not so black as it looks. Fig. 20 shows a yearling ram bred at Woburn Abbey by the Duke of Bedford.

Fig. 22 is the head of a ram from Duncansby Head, Caithness-shire, belonging to Mr. F. G. Sinclair of Barrogill Castle, and of a strain which he calls "Rocky," and which I believe to be a remnant of an aboriginal breed. (Cf. *Scottish Naturalist*, March 1912.)

"Spanish" or Piebald Sheep[2]

I am indebted to Mr. Heatley Noble of Temple Combe, near Henley-on-Thames, for a most interesting account of this breed,[3] from which I quote as follows :

"The origin of these sheep, often described as Spanish, Persian, African, Zulu, Syrian, Barbary, and even Merino,

[1] The head of a ewe bred in this park (Fig. 21), for which I am indebted to Mr. R. E. Holding, represents the best development of four horns that I have seen in the female sex.

[2] These sheep are called by various names—"Syrian," "Persian," "Zulu," "Barbary," "Jacobs," and "Spotted," but as the Syrian and Zulu origin are disproved by Mr. Noble, and the Spanish origin is doubtful, I think that Piebald or Spotted sheep is the best name for them.

[3] An effort to trace the history of the so-called Spanish Piebald sheep by Heatley Noble, published privately by the author, 1913.

seems to be entirely lost in antiquity, and although I have searched and written to many lands, my efforts to trace them have so far been without avail. There is a legend in the De Tabley family that their piebald sheep swam ashore from an Armada ship, and were afterwards brought to Tabley when Sir John Bryne of Queen's County sold his Irish estates and married an heiress of Tabley. There are pictures at Tabley painted about 1760 in which these sheep are shown. I am indebted to Lady Leighton-Warren of Tabley House for the above particulars, and also for two lithographs of very extraordinary heads (one of which is reproduced), taken by the then Lord de Tabley in 1822, and there is little doubt that the flock is one of the oldest in the country.

"There is some evidence that the pied sheep originally came to this country from the Spanish Peninsula, as the following will show. The late Sir Henry Dryden, writing in 1884, says: 'Many years ago a Spanish beggar woman came here (Canons Ashby) carrying a child on her back. I told her I could not speak Spanish. "But," she said, "there are many of my countrymen here." I made out that it was the *sheep*, and she explained that when the child saw the sheep it cried out that it recognised countrymen. I asked her more, and she said there were numbers of them where she lived, but I forget if I asked her the part of Spain.' The supposed Spanish origin of the Canons Ashby flock, will be referred to later.

"There is a flock at Charlecote Park belonging to Sir Fairfax Lucy, the ancestors of which are said to have come from Portugal. The following is an extract from a letter by Sir Fairfax Lucy, November 10th, 1910: 'I have looked up the sheep in a book of records we have, and all I can find about the sheep is this. A letter written by my grandfather, Lisbon, Rue d'Estrell, January 13th, 1756, in which it is stated, "Mr. Geo. Lucy remained in Portugal till June and brought with him the ancestors of the flock of white spotted sheep that graze in the Park amongst the deer."'

"Whatever may have been the case formerly, as far as I have been able to ascertain, these sheep are no longer found either in Spain or Portugal. I have a letter from the

Minister of Agriculture at Madrid in which he says they are unknown to him. I have been myself in the north and south of Spain, but notwithstanding numerous enquiries failed to find any trace of them.

"In a guide to the Domesticated Animals exhibited at the British Museum (Nat. Hist.), written by Mr. Lydekker, the following appears: 'South African Piebald Sheep. This breed appears to be originally a native of Zululand, but at least half-a-dozen flocks are kept in England.[1] Frequently the rams have only one pair of horns, and their colour is black with the exception of the face and the tip of the long tail, which are always white.' In regard to the native home of these sheep there appears to be much uncertainty amongst owners. It is, however, certain that they do not come from either St. Kilda or Uist. Perhaps the most satisfactory history exists in the case of the flock owned by Mrs. Farrer. The original parents of these sheep were brought home from the Cape about a century ago by the present owner's grandfather, Colonel Farrer, who believed that they had been imported into the Cape by Spanish or Portuguese settlers, who were supposed to have brought them from their own country. A portion of this original flock was given to Sir Henry Dryden's ancestor, so the Ingleborough and the Canons Ashby flocks have the same ancestors.[2] Lady Cowley's flock was imported at the time of the last Zulu War, about 25 years ago, and consisted of about thirty head. These were small, wholly black, two-horned sheep, with moderately long wool and long tails. A few of these black 'Zulus' were given by Lady Cowley to Mr. Lowndes, and these were subsequently crossed with piebald two-horned rams from the flocks of Mr. Whitaker and Sir Henry Dryden, with the result that the breed was greatly improved in size and stamina, whilst the rams frequently developed a second pair of horns. That the small black 'Zulus' and the larger piebald breed are identical, or nearly so, is rendered probable by the fact that the former are not unfrequently four-horned, and also from the circumstance that the Museum possesses

---

[1] There are about 50.—H. N.
[2] This is a mistake; see note on the two flocks.—H. N.

a head (presented in 1901 by the Rev. H. G. Morse) of a South African sheep which is black, with a white face, and has four horns. It is noteworthy that in this head the horns are much smaller than the English piebald rams, and also that the coat is short and hairy."[1]

I purchased a few of these "Zulu" sheep from Colonel Turnor of Pinkney Park in 1912, where they were called "Afghan" sheep. The photograph which he gave me (Fig. 23) of his old ram and some ewes shows that they are variable in colour, but all had long white tails, and most of them had white faces. On such good land as they were bred on they may do very well, but they were all very narrow and thin in make, and lost condition on my land. The lambs I bred from them by Sir C. Alexander's Piebald ram (Fig. 25) were much better than their dams, and followed the sire in colour.

With regard to the existence in North Africa of the Piebald sheep, I am informed by M. Geoffroy St. Hilaire of Tunis that the sheep of that province are all fat-tailed and have nothing in common with the Piebald sheep of which I sent him a photograph. These, he says, are much more like the sheep of the provinces of Constantine and Algiers, where, in the south, horned sheep are common; and he adds: "I think that one can only see in the specimens you show the descendants of sheep of which the variegated fleece and number of horns have particularly interested an amateur. The sheep have all the characters of the Barbary sheep; it is only their exceptional colour that makes one hesitate, and yet this mixed colour is not rare. As to the ewe with coarse open wool, called Spanish or Barbary, it is certainly a mountain sheep of Berberine origin. All the Berbers and Kabyles possess a variety identical with the representative which you possess."

Mr. R. Lydekker in *The Sheep and its Cousins*, p. 243, states that in his opinion the breed as we know it in England "is the result of crossing one or more of the small hairy black or piebald, and sometimes four-horned native

---

[1] This head is almost certainly that of a wether or very young ram. The coat on the face of our sheep is always short and hairy at any age.—H. N.

African breeds, with a woolly European strain"; but he does not bring any evidence to prove this, and the pictures of the breed at Tabley House painted about 1760, and by Stubbs at Wentworth about the same date, prove that the breed has been little if anything altered in appearance for at least 150 years.

When we consider their hardiness, freedom from lameness, prolific nature, and excellent mutton, as well as their remarkable appearance and beauty, it is surprising that they have received so little attention either from the owners of the parks which they have so long adorned or from farmers. Though naturally wilder than common low-country sheep, they are quite easy to tame; they are in my experience not inclined to jump or break out of ordinary enclosures. Their wool, though rather coarse, is capable of improvement when any care is taken in selection; and Mr. Noble tells me that he made last year 10d. and $10\frac{1}{2}$d. a pound for his clip unwashed. Though they do not, when kept on grass alone, fatten so quickly as some improved breeds, yet the wethers average about 60 lbs. dressed at 15 to 18 months old off grass which can only be described as second or third class.

They are excellent mothers, and produce more doubles than any breed which I have kept, and seem able to thrive on heavy and rather wet land with less attention than common sheep, though, judging from the various flocks I have seen, they are better kept on limestone or on chalk soils. I have tried crossing rams of the breed with ewes of many others, and the results have so far been very satisfactory. Fig. 24 is an old four-horned ram bred at Somerford Park. Fig. 25 is a shearling ram bred by Sir Claude Alexander from the same stock as the old ram in Fig. 2.

Fig. 26 is a ram lamb by the ram shown in Fig. 2, out of a fat-rumped ewe shown with the same lamb when four months old in Fig. 27. The other four lambs of this cross were all black, and showed the same curious formation of the tail (Fig. 28).

Fig. 3 shows twins bred in 1912 from a Wilts × Soay

ewe, by the same pied ram, and the one on the left is the only lamb I have bred from ewes not of the pied breed, whose colour was like that of the sire; for, as Mr. Noble says, "It is a curious fact that when our piebald sheep are crossed with any other domestic breed, the lambs are practically all born black with a white patch on the forehead and partly white tails."

The four-horned character is one which occurs more or less often in sheep in many parts of the world. It is common in some parts of Eastern Asia and occasional in North Africa and South America, but seems to be most characteristic, though not by any means a typical character, in the rams of the short-tailed North European and Iceland breed—called *Ovis polycerata* by Fitzinger—which exists in Shetland, the Hebrides, and the Isle of Man in various forms. The piebald breed seems to be the only long-tailed race in which the four-horned character is common, and I quite agree with Mr. Noble in the following remarks:

"It is doubtful whether the number of horns was originally two or four; when we say 'originally' we mean after the breed was established as a distinct domestic variety. On this subject and on the ancestors of sheep generally Professor Ewart very kindly sent the following interesting particulars—September 25th, 1911—'As there never existed, as far as we know, a wild four-horned sheep, it is safe to assume four-horned varieties are descended from two-horned ancestors, but from which wild ancestors four-horned sheep have sprung no man can tell. There were four-horned sheep in possession of the pile-dwellers of the Bronze Age. Our domestic sheep have probably all sprung from two wild Asiatic types, viz.: the Urial (*Ovis rignei*) and the Moufflon (*Ovis orientalis*). The Urial chiefly differs from the Moufflon in the face pits, they are larger and deeper; hence the Urial may have contributed to the making of some of the four-horned varieties.' Some of the flocks are and always have been two-horned, others again were once two-horned and are now four-horned owing to crossing and selection. There is no evidence to prove the Wentworth, Milton, De Tabley, and Abercairney flocks were not always more or less four-horned; on the other hand, we know the Canons Ashby, Ingleborough, and Chatsworth flocks were originally two-horned, but it seems probable the two former flocks have the same ancestry, also that they are of comparatively recent introduction. Personally, I am inclined to believe that they were all formerly two-horned, and that the second pair may have been the result of a cross with some Northern breed such as Hebridean or St. Kilda. Certain it is, that if a flock is allowed to breed in without the introduction of fresh blood the rams will revert to two horns, and after many years of carefully selecting none but four-horned rams to breed from, there are always a large percentage of ram lambs with only two horns. The ewes generally carry a single pair, but there are exceptions; some have four, even occasionally five; and in flocks where four-horned ewes have been especially selected the second pair are quite common. There are two distinct types of horns in the males; the most numerous are those in which one or both the upper pair grow straight up, turning backwards at the extremities, the lower pair curving round. In others both pairs turn down; in fact in some instances to such an extent they have to be cut back to prevent growing into the face. The colour of the horns is generally black, but there are numerous cases where they are light and parti-coloured."

The Fat-rumped Sheep

Another curious breed, little known in England, is the fat-rumped sheep, with a rudimentary tail consisting of only 6 vertebræ in one examined, and not reaching beyond the lobes of fat on the buttocks, which, when the sheep are in good condition, are of considerable size; it seems precisely similar, as Professor Ewart remarks, to the tail of the Manx cat.

The native country of these sheep is almost certainly some part of the Eastern Mediterranean, Tunis, Cyprus, or Syria, and some variety of this breed is the common sheep in many parts of North Africa and Western Asia. How long they have been in this country, and exactly where they came from, I have been unable to discover, but Sir C. Alexander bought them in Sussex, and their parents were imported fifteen or twenty years ago by the late Mr. Lucas, who brought them from Palestine in his yacht. They are a very hardy breed, more valued for their mutton than their wool, and a very similar breed has become quite popular in the United States, under the name of Tunisian sheep, on account of the good quality of their meat. As I was unable to procure a ram of the same or any kindred breed, I mated them with a large two-horned spotted ram (Fig. 2), and from five ewes got last year four black ewe lambs (Fig. 28) and the white ram (Fig. 26) which I show. The sire had a very long tail, and the crosses are in this respect nearer to the dam than the sire, and also follow their dams in having drooping ears, a character unknown in any British breed.

Last year I put my oldest ewe to an Old Wilts ram lamb, and the lamb (Fig. 29) is again nearer to the dam than to the sire. I sent two of the other ewes to Professor Ewart to be mated with an extraordinary long- and fat-tailed ram imported by the Marquess of Bute from Afghanistan (Figs. 30, 31), whose tail forms a great mass of fatty tissue nearly a foot in width and extending to the ground. I am not aware that this kind of sheep has been previously seen in England, but some crosses by him from Scotch Black-faced ewes have been bred as an experiment by Professor Ewart. I am indebted to Professor Ewart for Figs. 30, 31, 37, and 38.

I may add that the fat-tailed Afghan ram has a very coarse kempy wool outside and a thick coat of soft white wool below, as in the Kashmir goat and other animals inhabiting regions which are very cold in winter. Until he dies, and his tail can be dissected, we cannot tell how much flesh it contains; but an officer who had eaten the fat-tailed sheep kept by the Boers in some northern districts of the Orange Free State tells me that the tails, when cut off and roasted, contained enough lean meat to resemble a fat loin of mutton more than anything else, and that the rest of the carcass was so lean that it was not worth eating.

Barrow in his *Travels in the Interior of South Africa*, vol. i. p. 67 (1806), says : " . . . The broad-tailed breed of Southern Africa seems to be of a very inferior kind to those of Siberia and oriental Tartary : they are long-legged, small in the body, remarkably thin in the forequarters and across the ribs ; have very little intestine or net fat ; the whole of this animal substance being collected upon the hind part of the thigh, but particularly on the tail, which is short, broad, flat, naked on the under side, and seldom less in weight than five or six pounds ; sometimes more than a dozen pounds ; when melted, it retains the consistence of fat vegetable oils, and in this state it is frequently used as a substitute for butter, and for making soap by boiling it with the lie of the ashes of the salsola. This species of sheep is marked with every tint of color ; some are black, some brown, and others bay ; but the greatest number are spotted : their necks are small and extended, and their ears long and pendulous : they weigh from sixty to seventy pounds each when taken from their pasture ; but on their arrival at the Cape are reduced to about forty ; and they are sold to the butchers who collect them upon the spot for six or eight shillings apiece." He gives a portrait of a South African sheep, somewhat like but inferior to one which is stuffed in the Domestic Animals Gallery at the Natural History Museum.

The Welsh Sheep

The black sheep exhibited represents a strain of the Welsh mountain breed in which the colour has been fixed by long selection, and of which several pure-bred flocks are kept in North Wales and other places. The black colour is so persistent that every one of the pure lambs (nearly 100) which I have bred from them by a black Welsh ram have been pure black, and crosses made with Suffolk and Old Wilts rams on these ewes were also black.

Four Cheviot ewes put to the black Welsh ram have produced four black lambs and two black with white cap. This is a very curious fact, because I believe that the Welsh breeders have rejected black sheep for breeding as much as the Scotch and English have done, and I have certainly seen very few of this colour among ordinary Welsh flocks.

Low in his folio illustrations of the *Breeds of Domestic Animals of Great Britain*, published in 1842, figures a ram and a ewe from the mountains north of Neath, Glamorganshire, which are of the same chocolate-brown colour as the Manx "Loaghtan" sheep which I show, but with two horns and long tails.

I have seen no sheep of this colour in any part of Wales, and cannot hear of any that exist to-day, but it is interesting to know that the colour which is now only found in the short-tailed breeds of Man and Shetland may be found in the long-tailed breeds also.

I find the Welsh as hardy and as good mothers as any sheep I have. I have no trouble about jumping or breaking fences, and if properly handled by a shepherd who does not encourage fast dogs, they soon become as tame as any sheep. They are certainly able to live on grass which would starve English sheep, and though I am strongly opposed to the extreme starvation which is practised in the unstinted common hill-grazing of some parts of Wales, and which often leads to a heavy death-rate both among ewes and lambs, I have found that young sheep of this breed could winter successfully in my deer park on the same keep, with only a little hay in frost or snow, and that they would put on flesh in the spring as quick as the deer do.

If crossed with larger breeds in order to produce fat lambs, it certainly pays to keep the ewes on better land, and to give them some chaff and oats after lambing; and if the lambs are timed to fall about April 1st they grow fast and are appreciated by butchers in August and September, when real lamb is getting scarce. They cross well with many breeds; if fine wool is the object I should use a Shetland or a Ryeland; if early maturity is preferred, a Suffolk or Southdown would probably be the best ram to use. But though three- and four-year-old Welsh mutton is still heard of, I expect it is very seldom worth while to keep the pure breed over two years for mutton; and some of the Welsh carcasses in the butcher's classes at the Smithfield

Show seemed to me just as much overfed, and with as large a proportion of wasteful fat as any other breeds shown.

Fig. 32 is a ram by an imported Soay ram out of a black Welsh ewe, and might pass for a pure Welsh sheep if his shorter tail and finer wool, with less kemp in the breech, did not show the influence of the sire.

Fig. 33 is a horned Welsh ewe with her ram lamb as shown. I also show a shearling lamb which is a cross by a Welsh ram from a Cheviot ewe, which might pass for a pure Welsh, and am informed that a cross between the Cheviot ram and Welsh mountain ewe is found in some parts of Wales to be superior to the pure Welsh; and large quantities of them are bred in the neighbourhood of Brecon, where the mountain-grazing is not held in common.

Nothing can be more detrimental to good sheep-farming than the existence of large common grazings, as seen in Central Wales and in the Shetlands, and no real improvement, either to the land or the sheep, is possible where it exists. All the improvement which has been made in the Welsh breed, as shown at these shows and at Smithfield, has been done on enclosed land of a much higher quality and value than the common grazings, but the great difference between the average value of Welsh mountain and of Cheviot sheep is, I believe, much more due to careful breeding and management than to the land itself.

THE BLACK-FACED HIGHLAND SHEEP

The Black-faced Highland sheep is so well known that I need say little about the pure breed. The ewes that I show are brought to show the effect of crossing with other breeds. Fig. 34 shows a Black-faced ewe with her lamb by a Black Welsh ram. Fig 35, a ram by the one shown in Fig. 9, from a Black-faced ewe.

In England, as in its own country, the Black-faces are unrivalled for hardiness and for every good quality that a mountain sheep requires, except wool. I am not going to say anything about this, because Scotch breeders ought to know what pays them best better than any one else. But

I think that for the south the breed would be much better if the wool was shorter and not so coarse, and I should be inclined to shear lambs which are to be kept on low ground.

Many years ago I mated a lot of Black-faced ewes with a Cotswold ram in the hope of showing Scotch breeders of cross-bred sheep that the Cotswold was as good as the Wensleydale, which for many years was, I believe, the favourite cross. Though the Cotswold Black-faced cross was an excellent one as regards its ability to fatten at 16-18 months old on grass, the carcasses were too large and heavy to suit modern taste; two shearling wethers of the cross which I sent to Kelso Ram Fair weighed 102 and 104 lbs. dressed, which was a great deal heavier than the best judges supposed them to be when handling the sheep alive.

I commenced my new experiment with twenty ewe hoggs bought in May 1911, which endured the great heat and drought of that year as well as they have endured the extremely wet seasons of 1912 and 1913. None of them have ever been sick or sorry; they are wonderfully free from lameness; they lamb without help, and the lambs are always thriving.

Fig. 36 is from a photograph of a lamb's head with four horns which I found in the Royal Scottish Museum at Edinburgh, and which, as I am informed by Mr. W. Blackburn, was bred at Roshven, Lochailort, from "what seemed to be a small ordinary black-faced ewe with very black face and feet, which had been bought from a crofter near Mallaig." Whether the four horns are an entirely abnormal occurrence, or due, as Mr. Blackburn's shepherd suggested, to a previous meeting with a "St. Kilda" (? Hebridean) ram which had strayed from Fortwilliam, must remain doubtful, but I have never seen or heard of a pure breed Black-face sheep with similar horns.

The Siberian Sheep

The name "Siberian Sheep" has been provisionally adopted by Professor Ewart in a paper on "Domestic Sheep and their Wild Ancestors"[1] for a variety of uncertain origin, whose history we have as yet been unable to trace, and of which very few individuals exist in this country. In my opinion they are very nearly allied to and probably directly descended from the Shetland breed, or from a North European breed of very similar character. They are remarkable for their wool, which consists of an under-fur of very fine texture about 3 inches long, and an outer coat of stronger wool about 7 inches long. In colour they are variable, white, dark brown (almost black), but usually of a reddish brown, very like the Shetland "moorit" sheep. The face and parts of the head, neck, legs, and feet are sometimes white, and the tail of the Shetland type. Fig. 37 represents a ewe of this breed, with her two ram lambs, and Fig. 38 one of the same lambs when one year old. His own brother is sent for exhibition by Professor Ewart with some crosses which were bred this year at Fairslacks farm. Though it is as yet too soon to enable us to judge of their value, we hope to produce some very choice wool by breeding from this strain.

I have seen in a park in Gloucestershire some sheep sold by a dealer as "Siberian sheep," and have purchased a few of them, but these seem to have been crossed with some other breed, and are very variable both in wool and colour, though agreeing in their short tails with the Shetland breed.

One of these rams has horns unlike those of any domestic sheep I have seen, and showing a good deal of resemblance to those of the wild Urial of Asia (whose horns are shown in Fig. 39). This sheep had a very coarse hairy mane-like covering on his neck and throat, like that of the Urial or Moufflon, and is also remarkable for his thick deep body, but one of his half-brothers was a hornless sheep with very short soft and white wool. I have never seen or read of any sheep like these in Siberia, and cannot hear of any wool of similar character being known in the market from Siberia or elsewhere, and I shall be very grateful if any one can give me information which may enable me to trace the origin of the variety. Fig. 40 shows a cross between the Urial and

[1] *Trans. Highland and Agricultural Society of Scotland*, vol. xxv. (1913).

the Soay which is exhibited, as he appeared before shearing.

The Orkney Sheep

I gave in the *Annals of Scottish Natural History* some scanty notes on the indigenous sheep of Orkney, which seem to be nearly extinct except on the island of North Ronaldshay (Fig. 35*b*); and these, from what we know, appear to be hardly worth preserving as a distinct race.

But I have brought a curious sheep which I suppose to be descended from the Shetland breed, and is so very like the sheep figured by Low on his Plate I. as a ram of the ancient breed from the Isle of Enhallow that it has some interest as a throwback to a race now probably extinct.

Fig. 41 shows skull and horns of Marco Polo's wild sheep from the Pamir plateau in Central Asia. This breed has the largest horns of any sheep, though it inhabits one of the coldest regions at an elevation of 12,000 to 16,000 feet. The horns measure 69 inches round the curve and 50 from tip to tip.

Fig. 42 shows the horns of a breed known as "Wallachian," though I can find no evidence of its existence in that country at the present time, though it exists in Hungary. It is a sheep of the same type as the Scotch Black-faced, and has long, coarse wool, and horns of a very remarkable corkscrew shape which usually ascend at various angles, but sometimes, as in the specimen figured, are at right angles to the head. The horns shown measure $38\frac{1}{2}$ inches from tip to tip. A race of sheep with horns of somewhat similar character exists in North-Western China, but nowhere else, as far as I know.

Fig. 43 shows the stuffed head of the Argali wild sheep, shot in the Altai Mountains of Central Asia by myself. This breed is the largest and has the most massive horns of any sheep known, measuring, when fresh, as much as 20 inches in girth at the base and up to 62 inches round the curve.

For Figs. 16, 18, 22, and 35 I am indebted to the

editors of the *Scottish Naturalist*. The remainder, except where otherwise stated, were taken under many difficulties by Mr. J. Edwards at Colesborne.

In concluding these very imperfect notes, I wish to thank the Council of the Society for the facilities which they have afforded me in making this exhibition, and thus enabling me to call attention to a neglected branch of British agriculture.

FIG. 1.—Wether by Wilts Ram out of Soay Ewe.

FIG. 2.—Old Piebald Ram (sire of Lambs in Fig. 3).

FIG. 3.—Ewe by Wilts Ram out of Soay Ewe and her two Lambs by Piebald Ram (Fig. 2).

FIG. 4.—Ewe by Wilts Ram out of Soay Ewe and her two Lambs by Wilts Ram.

FIG. 5.—Wilts Ram Lambs bred by Mr. E. Berry.

FIG. 6.—Wilts Rams bred by Mr. E. Berry.

FIG. 7.—Shetland polled shearling Ram bred at Colesborne.

FIG. 8.—Yearling Shetland Rams bred at Colesborne.

Fig. 9.—Ram by Black-faced Ram out of Grey Shetland Ewe.

Fig. 10.—Hogg by White Shetland Ram out of Herdwick Ewe.

FIG. 11.—Head of four-horned Ram from Iceland.

FIG. 12.—Old Manx Ram bred by Mr. J. C. Bacon (sire of Fig. 13).

FIG. 13.—Yearling Manx Ram bred at Colesborne.

FIG. 14.—White two-years-old Manx Ram.

FIG. 15.—Manx Ewe and her five-horned Ram Lamb.

FIG. 16.—Hebridean Ram from North Uist.

FIG. 17.—Brown yearling Manx Ram.

FIG. 18.—Soay Ram, three years old.

FIG. 19.—Old Hebridean Ram.

FIG. 20.—Young Hebridean Ram bred at Woburn.

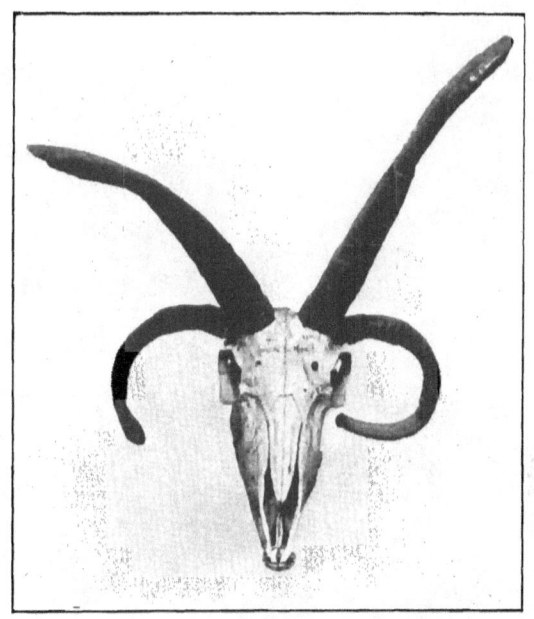

Fig. 21.—Horns of Hebridean Ewe bred at Ascott.

Fig. 22.—Rocky Ram from Caithness.

FIG. 23.—"Zulu" Ram and Ewes probably crossed by Piebald Ram.

FIG. 24.—Old Piebald Ram bred at Somerford Park.

FIG. 25.—Yearling Piebald Ram bred by Sir C. Alexander.

FIG. 26.—White Ram Lamb by Piebald Ram out of Fat-rumped Ewe.

FIG. 27.—Fat-rumped Ewe and her White Ram Lamb by Piebald Ram.

FIG. 28.—Fat-rumped Ewe with Ewe Lamb by Piebald Ram (Fig. 2).

FIG. 29.—Fat-rumped Ewe and young Lamb by Wilts Ram.

FIG. 30.—Tail of Fat-tailed Ram (Fig. 31) seen from behind.

FIG. 31.—Fat-tailed Ram from Afghanistan belonging to the Marquess of Bute.

FIG. 32.—Black Ram by Soay Ram out of Black Welsh Ewe.

FIG. 33.—Horned Black Welsh Ewe and her Ram Lamb.

FIG. 34.—Black-faced Ewe with her Lamb by Black Welsh Ram.

FIG. 35.—Four-horned Yearling Ram by Black-face × Shetland Ram (Fig. 9) out of Black-faced Ewe.

FIG. 35*b*.—North Ronaldshay Ram.

FIG. 36.—Black-faced Highland Lamb with four Horns.

FIG. 37.—"Siberian" Ewe and her two Lambs bred by Professor Ewart in Scotland.

FIG. 38.—"Siberian" Yearling Ram bred by Professor Ewart.

FIG. 39.—Horns of Wild Urial from N.W. India.

FIG. 40.—Hybrid Ram between Urial Ram and Soay Ewe bred by Sir C. Alexander.

FIG. 41.—Skull and Horns of Marco Polo's Wild Sheep.

FIG. 42.—Horns of the "Wallachian" or Zacketschaf.

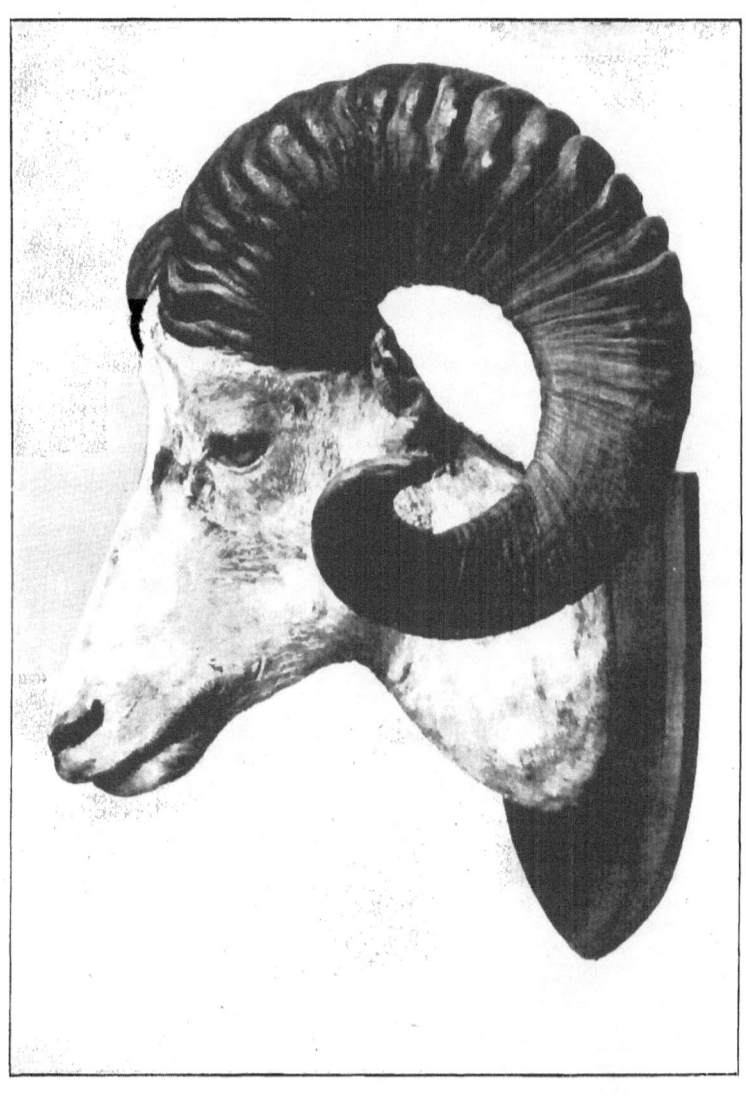

FIG. 43.—Head and Horns of the Argali from the Altai Mountains.

Current Status of Primitive and Rare Breeds

Rare Breeds Survival Trust Technical Consultant Mr Lawrence Alderson reviews the present positions of the breeds demonstrated at the 1913 Royal Show by Mr Elwes, and described by him in the book he wrote as a guide to the exhibits in his demonstration.

SHETLAND SHEEP

The natural variety in the colour of Shetland wool, as described by Elwes, continues to be a feature of the breed, but during the present century there has been strong selection for white wool. As a result of this policy some colours became very scarce, but it was unlikely that they would disappear completely, as being recessive to white, they can be carried from generation to generation without phenotypic expression.

Recently, the revival of self-sufficiency systems of farming and the growth of cottage industries has stimulated renewed interest in naturally coloured wool and has significantly increased the demand for sheep with a coloured fleece. The order of dominance of the colours is as follows: white (dominant), several types of grey, black, and moorit (recessive to all other colours). Thus white sheep may carry other colours as recessives, but moorit sheep will breed true for colour. Some sheep have a broken-coloured fleece.

Traditionally the natural-coloured wool formed the basis of the woollen industry of the Shetland Islands, which was famous throughout the World. Multi-coloured Fair Isle sweaters are knitted to patterns which have been handed down through families for generations. The hand made lace shawls made from the finest Shetland wool were so delicate that they could be drawn through a wedding ring.

A breed society was formed in the islands in 1926 to preserve and improve the breed. Most of the sheep on the islands are still of Shetland type, although Cheviot rams in particular have been introduced and have influenced the type of some sheep. In 1973, a register was opened by the Rare Breeds Survival Trust's Combined Flock Book for Shetland sheep on the mainland.

The Shetland is a hardy breed, and is adapted to tolerate the wet, exposed conditions of its native islands. It is an alert, active animal and is not so amenable to shepherding with dogs as many other breeds. It is a small animal, and adult ewes weigh 60 to 70lb. Its tail is short and fluke-shaped, being broad at the base and tapering to a fine point barely covered with wool. Rams are usually horned and ewes polled.

The wool is the finest produced by any native British sheep, and has a Bradford count of 54-60. The staple length is 4in and the fleece weight is 2½lb. The Shetland has also demonstrated its value as a hill ewe producing prime lamb, and trials carried out by the North of Scotland College of Agriculture have shown it to be more efficient than the Scottish Blackface.

Manx Loghtan Sheep

When Elwes described the Manx in 1913 he reported that both white and black specimens were found. Since that time 'moorit' has been established as the correct colour for the breed. This process has been achieved with relative ease because moorit breeds true, but some colour variations still occur. In particular, some animals have white markings, usually appearing as a spot on the poll, or as a blaze, or on the tip of the tail, or on the legs, but on some animals the white marks may be more extensive and occasionally appear in the fleece. Other animals show a faint 'Soay' pattern (lighter coloured rump patch and underparts) and this probably traces back to the introduction of a Soay ram to the Isle of Man after the second World War.

Elwes reported that Manx sheep increased in size and improved their productivity when brought down to better land, and this still holds true. Mature ewes on the mainland weight about 80 to 90lb. Emphasis in breeding programmes is still given to the selection of animals with four good horns, and the illustrations used by Elwes show that even then a sound horn structure was important, with the top pair growing upright and the lower pair not impeding the movement of the jaw.

The breed has increased numerically in the last decade, and in 1982 there were 180 ewe lambs registered in addition to unregistered sheep on the Isle of Man. Flocks are now located throughout the United Kingdom.

Soay Sheep

Elwes considered that Soays had little, if any, value from an agricultural point of view, although their wool was very soft and fine and they were long lived and prolific. He stated that they seemed to require good feeding and plenty of room. This evaluation bears little resemblance to the Soay sheep that we know today. It is possible that the sheep on which he based his evaluation were very inbred, as their progeny from crosses with other breeds performed very well.

In 1932, the Marquess of Bute transferred 20 rams, 44 ewes, 22 ram lambs and 21 ewe lambs from Soay to Hirta, when the human population was evacuated from the latter island. The modern mainland population is descended from these sheep. While they do not fit into the normal patterns of sheep production, they do possess qualities which are of commercial value. In particular they produce a lean carcase at any weight, and the lean-to-fat ratio is better than that of any other breed. When its production is measured in relation to its body weight it is more efficient than many other breeds. A Soay ewe rearing twins will yield 1.57 times her own body weight of milk in one lactation, compared with only 1.21 times for the Scotch Halfbred.

Soay sheep rarely have difficulty at parturition, and their thrifty nature makes them an 'easy-care' sheep, a factor further enhanced by their naturally shed fleeces. They have been used successfully to reclaim the waste sand tips from china clay works in Cornwall. Ewes on the mainland weigh about 55lb and yield a fleece weighing about ¾lb. Although pure bred lambs gain only about ⅓lb per day, crossing with a Down ram can double the growth rate of the lambs.

The feral population on Hirta fluctuates considerably between about 500 and 1,000 breeding animals, but a significant population has been established on the mainland and about 200 ewe lambs are registered each year.

Hebridean Sheep

It would seem that the Hebridean has been a cause of confusion for some time. Recently its name was changed from St Kilda to Hebridean in order to avoid confusion with the Soay sheep which inhabit the St Kilda islands. Elwes made the same point: "The Hebridean breed is perhaps more generally known under the name of St Kilda sheep . . . but . . . they are not the pure old St Kilda breed . . .". He also observes that the breed is impure, as there is no fixed type.

Since that time a much greater degree of uniformity has been achieved throughout the breed, but some variations still occur. The fleece colour is black, but there are two types. The first has a high lustre and is darker with a stronger, longer staple. It often turns silvery with age, as grey fibres appear in the fleece. The second type has a brown tinge and is softer and finer. The legs of both types are black. White marks may occur, usually in the shape of a white spot on the poll, and occasionally more extensively, but are not acceptable under the current breed standards.

Apart from their colour, Hebrideans are very similar to Manx Loghtans. They are small and fine-boned. The tail is short, usually reaching to just above the hocks. A defect, known as 'split eyelid' occurs and seems to be linked to the multi-horned factor, as it rarely, if ever, is seen in two-horned animals. The four-horned gene is dominant to the two-horned. One of the four-horned Hebridean rams illustrated by Elwes had his top pair of horns curling forward almost into his face, but the shape and structure of the horns of this breed have now been much improved, and the occasional use of two-horned rams in breeding programmes helps to strengthen the horns.

Traditionally, the breed was kept in parks adjacent to country houses, and this still accounts for a large proportion of the breed. About 200 females are registered each year.

Although today, Hebrideans are still mainly kept in parks,

some breeders have started to use them in crossing programmes with commercial breeds, and have reported encouraging first results.

North Ronaldsay Sheep

Elwes allows the North Ronaldsay sheep only one paragraph, and declares them to be "hardly worth preserving as a distinct race".

The majority of the breed remains on its native island and is administered by a Sheep Court. In 1973, the Rare Breeds Survival Trust took a group of these sheep to Linga Holm, a small island close to Stronsay in the Orkney archipelago, and these have been established as a reserve population comprising 25 rams and 175 ewes. At the same time a few small units were established on the mainland.

The North Ronaldsay is a small sheep, adult ewes weighing about 55lb. They are fine-boned and short-tailed. The rams are horned, but the ewes may be either horned or polled. The colour of the face and legs is variable, and the colour of the fleece shows the same range as in the Shetland breed, but with a greater mixture of colours. White and grey are the most common colours: black fibres in the fleece tend to be stronger and coarser.

The most distinctive characteristic of the breed is its ability to survive on a diet of seaweed. On its native island it is excluded from the grassland by a wall which encircles the cultivated areas. The sheep have adapted physiologically to this diet, and as a result find difficulty in adjusting to a normal sheep diet, and many succumb to copper toxicity.

Jacob and Black Welsh Mountain Sheep

Since Elwes included these breeds in his account of primitive breeds, they have been improved significantly and have attracted the attention of commercial flock masters. They have taken their place in the pattern of sheep production in Great Britain and are no longer relic breeds.

Norfolk Horn Sheep

Elwes recorded that very few of the breed remained in 1913, and for most of the present century it has existed under the threat of imminent extinction. Currently no pure bred Norfolk Horn sheep survive. At the end of the First World War, only two flocks remained, owned by Mr J. D. Sayer and Col McCalmont, and the latter flock was dispersed in 1919. Thus it is due entirely to the efforts of Mr Sayer that the Norfolk Horn survived. Its history from 1919 to 1969 was described by Mr Frank Rayns as follows: "From 1919 onwards, J. D. Sayer owned the whole breed. In 1930 they were down to 20 ewes besides rams, lambs and wethers; in all probability 50 to 60 sheep remained in 1930. In 1948, there was a breed crisis — only two rams were left and both were semi-cryptorchids. After temporary lodgement at the Norfolk Agricultural Station, one of these passed with two ewes to Dr (later Sir John) Hammond at the Institute of Animal Pathology, Cambridge, where the inheritance of cryptorchidy was being studied. Fortunately both rams proved fertile despite their abnormality, and so maintained the purity of the breed. One of their male offspring was returned to J. D. Sayer, who in 1950 had two ewes and two rams, and John Hammond's number had increased to six ewes and three rams. The Norfolk breed then totalled 13 animals.

"Before Mr J. D. Sayer's death, the Cambridge sheep were returned to him and at the time of his death (1954) his relation by marriage Mr R. E. Smith, Wordwell Hall, Culford, Suffolk, had taken all that remained — two ewes, two rams and two shearlings. Between 1954 and 1959 the flocks were again dispersed. Half remained at Wordwell Hall, half went to Mr W. Harvey, Howe Hall, Little Bury Green, Saffron Waldron, who is also distantly related to Mr Sayer by marriage. In 1959, both lots of sheep were given to the Zoological Society of London and lodged at Whipsnade Zoo. In 1965, Whipsnade had six ewes and seven rams.

"In 1968, at the suggestion of Sir Peter Greenwell, of Woodbridge, Suffolk, all the remaining Norfolk Horns were transferred to the National Agricultural Centre at Stoneleigh, and there the surviving six ewes, six rams and eight crossbred Suffolk ewes were lodged. Unfortunately the numbers had fallen to three rams and three ewes by April 1969, and the results of inbreeding became manifest in the birth of spastic lambs: but mated with the Suffolk, the progeny are normal and vigorous and by this means both the type and potential of the old breed may be retained."

The breeding back programme, commenced at Whipsnade and continued at Stoneleigh, has been instrumental in saving the 'breed'. The Suffolk was used in the programme as it was evolved, by crossing the Norfolk Horn with the Southdown, and thus is the breed most likely to have retained Norfolk Horn genes. This programme has now been extended and includes eight breeders who participate in a group breeding plan designed to fix the characteristics of the breed while limiting the level of inbreeding. Numbers are still very low, and in 1982 only 53 ewe lambs of varying degrees of purity were registered.

Wiltshire Horn Sheep

The status of the Wiltshire Horn has changed very little since Elwes described the breed in the early years of this century. It remains located mainly in Northamptonshire and Anglesey, where the official breed sales are held. Recently flocks have been established near its native county.

The Wiltshire Horn is unique among British breeds of sheep. It grows a thick, hairy, matted fleece which peels off in the early summer. This characteristic eliminates the need for shearing and dagging, and is claimed to reduce the incidence of fly strike and other ectoparasites. A breed society was formed in 1923, and despite a decline in the fortunes of the breed in the 1950s and 1960s, there are now about 50 flocks and 1,500 registered ewes. The breed is used mainly as a sire of fat lamb, and transmits high growth rate and a good killing-out percentage to its progeny.

Adult rams weigh up to 300lb and adult ewes approximately 160 to 170lb. Both sexes are horned, with white legs and faces.

Other Breeds of Sheep

Elwes included notes on the Scottish Blackface, Siberian and Fat-rumped sheep in his account of primitive and rare breeds. The Scottish Blackface is the most popular breed of sheep in Britain, while the Siberian and Fat-rumped sheep are no longer to be found here, and are not relevant at this time.

However, Elwes made no mention of several other breeds which are relevant, namely the Castlemilk Moorit, the Boreray, the Portland, and the Whitefaced Woodland.

The Castlemilk Moorit is a relatively new breed which was created by the Buchanan-Jardine family on their estate in Dumfriesshire in the early years of the present century. In colour it is similar to the Soay, but it is larger. The population is very small.

The Boreray exists mainly as a feral flock of 350 to 450 animals on the island of Boreray in the St Kilda group. It is a small sheep with adult ewes weighing about 60lb. Both sexes are horned. The legs and face are mainly black-and-white or grey-and-white. The tail is short, and the fleece is shed naturally during the summer.

The Portland is a native of South-West England and probably is closely related to Welsh Mountain sheep. The wool of newborn lambs is foxy-red, changing to white within a few months, but the face and legs remain tan in colour throughout life. Both sexes are horned and the horns of the ram are strong and spiralled. Adult ewes weigh 70 to 75lb and are clean-legged and fine-boned. The wool is fine and close. The breed is noted for the flavour of its meat and its poor prolificacy. In 1982 there were 46 registrations for pure bred ewe lambs.

The Whitefaced Woodland is one of the largest hill sheep and is a native of the southern Pennines. Mature ewes weigh between 135 and 160lb, and are strong-boned and robust. The face and legs are white, and both sexes are strongly horned. The wool is white and relatively fine for a hill breed. The Whitefaced Woodland is capable of competing on equal terms with other hill breeds. About 180 ewe lambs are registered each year in addition to unregistered sheep in the breed's area of origin.

Shetland ram

Four-horned Manx Loghtan ram

Soay sheep

Four-horned and two-horned Hebridean sheep

Boreray ram

Portland ewe and lamb

Norfolk Horn sheep

Castlemilk Moorit sheep

www.ingramcontent.com/pod-product-compliance
Lightning Source LLC
Chambersburg PA
CBHW032006220426
43664CB00005B/164